Grades K-4

Biology

Laboratory Notebook

3rd Edition

Rebecca W. Keller, PhD

# Real Science-4-Kids

Cover design: David Keller
Opening page: David Keller, Rebecca W. Keller, PhD
Illustrations: Janet Moneymaker and Rebecca W. Keller, PhD

Copyright © 2019 Rebecca Woodbury, Ph.D.

All rights reserved. No part of this publication may be reproduced, stored in a retrieval system, or transmitted, in any form or by any means, electronic, mechanical, photocopying, recording, or otherwise, without prior written permission from the publisher. No part of this book may be reproduced in any manner whatsoever without written permission.

Focus On Elementary Biology Laboratory Notebook—3rd Edition
ISBN 978-1-941181-34-8

Published by Gravitas Publications Inc.
www.gravitaspublications.com
www.realscience4kids.com

A Note From the Author

Hi!

In this curriculum you are going to learn the first step of the scientific method:

Making good observations!

In the science of biology, making good observations is very important.

Each experiment in this notebook has several different sections. In the section called *Observe It*, you will be asked to make observations. In the *Think About It* section you will answer questions. There is a section called *What Did You Discover?* where you will write down or draw what you observed from the experiment. And finally, in the section *Why?* you will learn about the reasons why you may have observed certain things during your experiment.

These experiments will help you learn the first step of the scientific method and.....they're lots of fun!

Enjoy!

Rebecca W. Keller, PhD

Contents

Experiment 1:	WHAT IS LIFE?	1
Experiment 2:	TAKING NOTES	9
Experiment 3:	WHERE DOES IT GO?	23
Experiment 4:	WHAT DO YOU NEED?	38
Experiment 5:	YUMMY YOGURT	48
Experiment 6:	LITTLE CREATURES MOVE	53
Experiment 7:	LITTLE CREATURES EAT	65
Experiment 8:	OLDY MOLDY	77
Experiment 9:	NATURE WALK: OBSERVING PLANTS	85
Experiment 10:	WHO NEEDS LIGHT?	92
Experiment 11:	THIRSTY FLOWERS	105
Experiment 12:	GROWING SEEDS	116
Experiment 13:	NATURE WALK: OBSERVING ANIMALS	129
Experiment 14:	RED LIGHT, GREEN LIGHT	136
Experiment 15:	BUTTERFLIES FLUTTER BY	148
Experiment 16:	TADPOLES TO FROGS	160

Experiment 1

What Is Life?

I. Think About It

Biology is the study of life. In this experiment you will explore the differences between living things and non-living things.

❶ What do you think the differences are between life and non-life?

_____
_____
_____
_____
_____
_____
_____

❷ What do you think makes you different from a rock?

_____
_____
_____
_____
_____
_____
_____

❸ What do you think makes a frog different from a table?

_____
_____
_____
_____
_____
_____
_____
_____

❹ Do you think a rock ever dies? Why or why not?

_____
_____
_____
_____
_____
_____
_____
_____

II. Observe It

Find one living thing and one non-living thing to observe. Write the name of each thing in the space provided on the next page. Use the following questions to help with your observations.

- Can the item move?
- Does the item breathe?
- Does the item consume food?
- Can the item reproduce itself?
- What else can you notice?

Write or draw your observations in the spaces on the next page.

Experiment 1: What Is Life?

	Living Thing	Non-Living Thing
Write	_____	_____
	_____	_____
	_____	_____
	_____	_____
	_____	_____
	_____	_____
	_____	_____
	_____	_____

	Living Thing	Non-Living Thing
Draw	_____	_____

III. What Did You Discover?

❶ List four things that are different between living things and non-living things.

_____
_____
_____
_____
_____

❷ If you traveled to a faraway planet, how would you know which things were alive and which were not alive?

_____
_____
_____
_____
_____

❸ Write your own definition of life.

_____
_____
_____
_____
_____

IV. Why?

Biology is the study of life. The first step in studying life is knowing what is alive and what is not alive. It can be very easy to know the difference between something that is alive and something that is not alive. However, defining life can be challenging, even for scientists. In general, living things consume some sort of food source, reproduce themselves, and respond to their environment.

V. Just For Fun

Imagine that you have traveled to a faraway planet that no one has ever been to before. Using your imagination, think about the kinds of living things you might discover there. Think of as many details as you can, and then write about or draw pictures of these imagined living things on the faraway planet.

What would you name this planet? Do the imaginary creatures have names? Record your ideas on the next page.

Living Things on Planet _____

Experiment 2

Taking Notes

Introduction

The most important tool a biologist can use is the ability to make good observations. In this experiment you will practice making observations of the things around you.

I. Think About It

What does an ant look like? What is inside a tree? What is the shape of a bird wing? What is the color of a rose? Think of 5 questions about biology that you are interested in and write them in the following boxes. Then think about possible answers. Draw or write your ideas in the boxes below your questions.

Question 1: _____
_____
_____

Question 2:

Question 3: _____
_____
_____

Experiment 2: Taking Notes 13

Question 4:

Question 5:

II. Observe It

Take your *Laboratory Notebook*, a pen or pencil, and a magnifying glass, and go outside. Leave all electronic devices inside. Do not use a camera, smartphone, tablet notebook, or recording device to record what you observe. Rely only on your five senses.

Find an open space, park, or just an area in your back yard. Sit, listen, and observe the things around you for 10 minutes. What do you see? What do you hear? What colors are visible? What animals are nearby? What plants can you observe? Carefully draw and write what you observe. Notice things such as colors, textures, and shapes. Use the magnifying glass to view any small details that interest you.

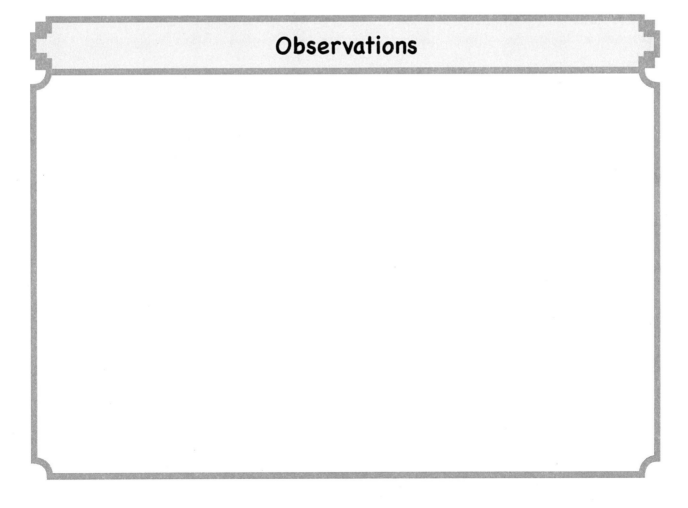

Observations

Observations

Experiment 2: Taking Notes 17

Observations

III. What Did You Discover?

❶ What did you hear?

_____
_____
_____
_____
_____

❷ What colors did you see?

_____
_____
_____
_____
_____

❸ Was the surrounding light bright or dull?

_____
_____

❹ What animals did you observe?

_____
_____
_____
_____

❺ Were there plants? If so, what kinds?

_____
_____
_____
_____
_____

❻ How much of what you saw was "new" for you?

_____
_____
_____
_____
_____

❼ What is the difference between a "natural object" and one designed and made by people?

_____
_____
_____
_____
_____
_____
_____

IV. Why?

Observing the natural world is how biologists started learning about plants and animals. Without the skill of good observation, very little about the world would be known. The more you look, the more you see. When we first look at something, we tend to form an impression about it and think we know all about it. But we often miss the details. Sitting quietly and using the five senses and simple tools such as a magnifying glass allows us to see more details, hear more sounds, and make better observations.

V. Just For Fun

Go someplace you have never been before, like a park, museum, library, or ice-cream parlor. Walk around with your *Laboratory Notebook* and record what you see, hear, and smell—or taste, if you're at an ice cream parlor (try a new flavor you've never had before!).

Observations of _____

More Observations of _____

More Observations of _____

Experiment 3

Where Does It Go?

I. Observe It

❶ Collect some different objects to observe.

❷ Look carefully at the objects and make observations about them.

❸ In the spaces provided, name each object and describe each in detail using words or pictures.

Experiment 3: Where Does It Go?

II. Think About It

❶ Look at all of the objects you described. Think about different groups you might use to sort them. You might use small or round or white or fuzzy.

❷ Name five groups you will use to sort the objects. Put the name of each group in the gray box. Put each object in **ONE** group only.

❸ Are there objects that fit in more than one group? If so, re-sort as many objects as you can into new groups.

___	___	___	___	___
___	___	___	___	___
___	___	___	___	___
___	___	___	___	___
___	___	___	___	___
___	___	___	___	___

❹ Can you do it again?

___	___	___	___	___
___	___	___	___	___
___	___	___	___	___
___	___	___	___	___
___	___	___	___	___
___	___	___	___	___

III. What Did You Discover?

❶ What did you observe about the objects you collected?

_____
_____
_____
_____
_____

❷ Was it easy to pick groups to sort the objects? Why or why not?

_____
_____
_____
_____
_____

❸ Was it easy to decide which objects would go in each group? Why or why not?

_____
_____
_____
_____
_____

❹ The objects in a group have the same feature (for example, round or small). List some features that were different between objects in the same group.

The _____ objects
were all _____ but some
were also _____.

The _____ objects
were all _____ but some
were also _____.

The _____ objects
were all _____ but some
were also _____.

The _____ objects
were all _____ but some
were also _____.

The _____ objects
were all _____ but some
were also _____.

IV. Why?

It can be hard to sort objects into groups. Some round objects may also be fuzzy, like a cotton ball. And some other round objects might be smooth like a rubber ball. Some smooth objects might also be large. And some smooth objects might also be small. How do you decide which object to put in which group?

This can be a difficult problem, even for scientists. Living things have lots of different features, and it can be hard to figure out which living things go in which groups. Do you sort all the green creatures in one group and all the brown creatures in another group? This would be one way to sort green grass and bears. But what about a tree? A tree is both green and brown. Does a tree go with the grass or with the bears?

Scientists sometimes discover a new living thing—a creature they have never seen before. The first thing a scientist does is make careful observations about the creature. Is it green or gray? Does it have smooth skin or scaly skin? Does it live in the water, or does it live on land? Does it eat vegetables, or does it eat other animals? Can you see it with your eyes, or do you need to use a microscope to see it?

All of these observations help scientists know which group a new creature should go into. By putting it into a group, scientists can better understand what is the same and what is different about the new creature compared to other creatures.

V. Just For Fun

You are an explorer traveling to places on Earth where no one has been before. In one of these remote areas, you find a new creature that no one knows about. It has the following features.

- It is green.
- It eats flies.
- It lives in trees.
- It flies with wings.

If you had to put your new creature into a group, would you group it with frogs, monkeys, or butterflies?

Draw a picture of this new creature.

The scientist who discovers a new living thing gets to name it. What would you name this creature you found?

A New Creature

Experiment 4

What Do You Need?

I. Observe It

❶ Cells are like little cities. There are lots of jobs that need to be done by lots of different workers inside a city, and there are lots of jobs that need to be done inside a cell.

Your mom and dad also do lots of different jobs at your house. Observe some jobs that your parent does during the day.

List some jobs that you observe your mom or dad doing at your house.

Job	
Job	
Job	
Job	
Job	
Job	
Job	

❷ Pick one of the jobs you listed, and think about all the tools or items you think your mom or dad would need to have when doing this job.

Write the name of the job and list or draw the items you think would be needed to do the job.

Job	
Items Needed	

❸ Draw a picture showing your mom or dad doing their job with the items needed to do it.

Job _____

❹ Pick one of the items your mom or dad uses to do their job. Draw a picture of the item, showing details.

Item _____

Job _____

Experiment 4: What Do You Need? 43

II. Think About It

For the item you drew, answer the following questions verbally or in writing.

❶ What is the item?

_____

❷ Where did the item come from?

_____

_____

❸ How did the item get there?

_____

_____

❹ Who made the item?

_____

_____

❺ What is the item made of?

_____

_____

❻ Where does the material that the item is made of come from?

_____

_____

III. What Did You Discover?

❶ Think about all the different jobs that your mom or dad does in your house. Make a list of these jobs.

_____
_____
_____
_____
_____

❷ How many items does your mom or dad need to have when doing these jobs? List these items.

_____
_____
_____
_____
_____

❸ How many people does it take to make the item you drew in Step ❹ of *I. Observe It*? List as many as you can.

_____
_____
_____
_____
_____

❹ If your mom or dad had to make their own item to do the job, how many more jobs would they need to do? List a few.

_____

_____

_____

_____

❺ If your mom or dad had to make all of the items in your house, how many jobs do you think they would have? List a few.

_____

_____

_____

_____

❻ Do you think it helps to have a city that can make certain items so your mom and dad can buy the items they need to have in order to do other jobs?

_____

_____

_____

_____

IV. Why?

Think about all of the items your mom and dad need to have when doing certain jobs in your house. Notice that they need lots of different items because there are lots of different jobs. It takes lots of different people doing lots of different jobs in different parts of the city or country to make the items your mom or dad uses.

A cell works in much the same way as a city works. There are places in the cell where different jobs get done, and there are lots of different types of molecules that are needed for a cell to do those jobs. In order for a cell to live, a cell must make many of these molecules itself. In a cell, there are many molecules doing many different jobs to make all of the molecules a cell needs.

Each part of a cell has a different job to do. And each job a cell does needs different molecules. A cell must make sure all of the molecules it needs for living are in the right places at the right time and in the right amount.

When a cell does not have all the molecules it needs to do all the jobs it has to do, or when a cell does not have enough molecules, or when the jobs are not done in the right way, the cell cannot live. In a similar way, a city would not work if people were not doing the right jobs, or were not in the right place at the right time, or if the jobs were not done in the right way.

V. Just For Fun

Observe one of the items your mom or dad uses to do a job.

- What materials is it made of?

- How many different materials can you observe?

- Choose one of these materials and look up the way it is made. Write about or draw what you find out.

Job _____

Experiment 5

Yummy Yogurt

I. Think About It

What do you think the differences are between milk and yogurt? Write your answer below.

_____
_____
_____
_____
_____
_____
_____
_____
_____
_____
_____
_____
_____
_____
_____
_____
_____

II. Observe It

In the space below, write down the differences you observe between milk and yogurt.

Milk	Yogurt
_____	_____
_____	_____
_____	_____
_____	_____
_____	_____
_____	_____
_____	_____
_____	_____
_____	_____
_____	_____
_____	_____
_____	_____
_____	_____

III. What Did You Discover?

❶ How did the yogurt smell?

_____
_____
_____

❷ Was the yogurt thicker than the milk?

_____
_____
_____

❸ What did the yogurt taste like?

_____
_____
_____

❹ How did the yogurt feel in your hands? Did it feel different from the milk? Why or why not?

_____
_____
_____
_____

IV. Why?

Although yogurt is made from milk, yogurt looks, feels, tastes, and smells different from regular milk.

When bacteria are added to milk, the bacteria produce lactic acid. This causes changes to the milk that turn the milk into yogurt. These changes make yogurt thicker than regular milk and give yogurt a particular smell and taste.

When you eat yogurt that contains "live and active cultures," you are eating live bacteria. The bacteria found in yogurt are bacteria that are good for you. They help you digest the yogurt.

V. Just For Fun

Try making your own yogurt mixtures. What other foods could you add to the yogurt to change its taste or color?

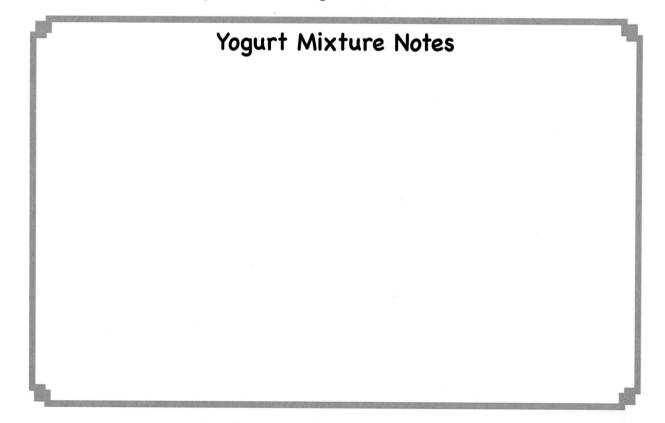

Yogurt Mixture Notes

Experiment 6

Little Creatures Move

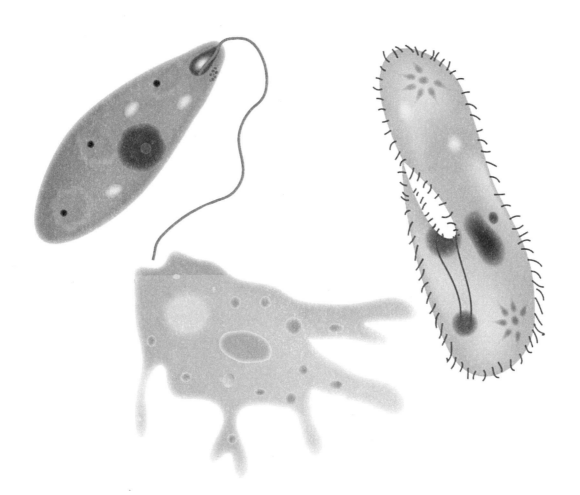

Introduction

In this experiment you will take a look at some tiny creatures that are too small to see when you use only your eyes.

I. Think About It

If you look at some pond water with a microscope, what do you think you will see? Draw what you think you will see.

Experiment 6: Little Creatures Move   55

II. Observe It

❶ Take some pond water and put it under the microscope. Draw what you see.

❷ See if you can observe moving creatures that are different from each other. Draw one here.

❸ **Draw a different moving creature here.**

❹ Draw a different moving creature here. Note if it moves differently.

❺ Draw a different moving creature here. Note if it moves differently.

❻ Are there two creatures that are similar? Draw them here.

❼ Are there two creatures that are different? Draw them here.

III. What Did You Discover?

❶ Did the pond water look like you thought it would? Why or why not?

_____

_____

❷ What was the first thing you noticed about the pond water?

_____

_____

❸ Was there anything you did not expect to find in the pond water? Describe it.

_____

_____

❹ How many different creatures did you find?

_____

_____

❺ Did you find two or more that moved in different ways?

_____

_____

❻ Describe your favorite creature. Explain why it is your favorite.

_____

_____

_____

IV. Why?

Pond water is full of little creatures. In fact, little creatures are found in soil, in hay, and in oceans and rivers. You might also find unwanted creatures on your toothbrush! Many of these little creatures are called protists (also called protozoa).

There are many different kinds of protists. Many types of protists move by using a large whip-like tail called a flagellum. Other types of protists use small hairs called cilia to move. And still other types of protists crawl using "false feet."

Using a microscope, you can observe protists move. You might see them move forward and backward. You might see them bump into a piece of food or even bump into each other. You can see them roll and stop and turn and then start moving again.

Protists need to move to find food or escape from danger or find a place to rest, just like you do. Humans have legs to move. If your mom calls you for dinner, you need to get up from reading this book and walk to the table to eat. If you are strolling in the park and a big dog starts barking, you might want to run away as fast as your legs will move. When it gets dark and you are ready for bed, you walk to your room (after using your new protist-free toothbrush on your teeth) to go to sleep. You are designed differently from a protist, but protists can also use their bodies to move just like you do!

V. Just For Fun

Leeuwenhoek looked at organisms in his mouth. What's in your mouth?

Spit onto a slide and see if you can observe any moving things in your saliva. Draw anything you discover.

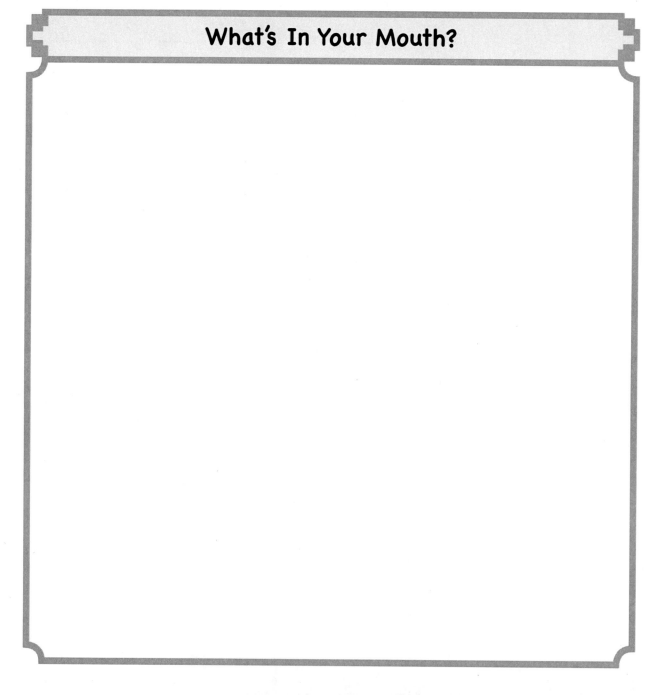

What's In Your Mouth?

Experiment 7

Little Creatures Eat

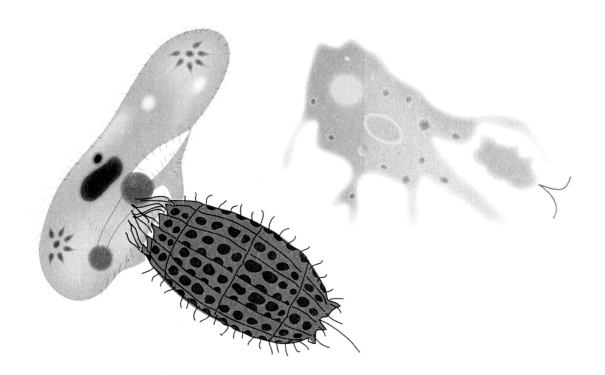

Introduction

In this experiment you will take another look at very tiny creatures and observe how they eat.

I. Think About It

If you use a microscope to look at a protist eating, what do you think you will see? Draw what you think you will see.

II. Observe It

❶ Take some pond water and put it under the microscope. Observe whether any protists are eating. Draw what you see.

❷ See if you can observe different creatures that are eating. Draw one below.

❸ Draw a different creature that is eating.

❹ Draw another different creature that is eating.

Experiment 7: Little Creatures Eat 71

❺ Draw the food one creature might be eating.

❻ Are there two creatures that are eating in the same way? Draw them below.

Experiment 7: Little Creatures Eat 73

❼ Are there two creatures that are eating in different ways? Draw them below.

III. What Did You Discover?

❶ Did the protists eat like you thought they would? Why or why not?

_____
_____

❷ What was the first thing you noticed about the eating protists?

_____
_____

❸ Was there anything you did not expect to find while you were watching the protists eat? Describe it.

_____
_____

❹ How many different ways did they eat?

_____
_____

❺ Describe your favorite creature. Explain why it is your favorite.

_____
_____
_____
_____

IV. Why?

Protists eat in different ways. Some protists make their own food, like the green euglena does. Some protists use their tiny hairs to sweep food into their mouths, like the paramecium does. And other protists capture their food with their feet, like the amoeba does. Because there are lots of different kinds of protists, there are lots of different ways protists eat.

Protists can eat lots of different kinds of food. They can eat algae or other small plants. They can eat yeast, and they can eat other protists. Imagine what might happen when two protist-eating protists meet. Who gets to eat whom?

You also eat, but like most humans, you usually use your mouth to eat. Humans can eat both plants and animals. You usually don't need to hunt for your food—unless your brother steals your piece of cake! But you do need food. You are not like a euglena who can make its own food, and you can't catch your food with your feet like an amoeba. In a city, you need to rely on other people who can provide you with the food you need. You need milk from the dairy and bread from the bakery and eggs from the farm and just for fun—chocolate from the chocolate factory! A protist doesn't have other protists finding food for it—it must find its own food.

V. Just For Fun

Do protists like chocolate? Try it. Place a small piece of chocolate on the slide and see if protists eat it. Record your observations.

Protists and Chocolate

Experiment 8

Oldy Moldy

Introduction

Grocers work hard to sell good fruits and vegetables. They throw away many pounds of moldy fruit each year. In this experiment, we will find out how picky fungi are. Will a particular fruit fungus grow on bread?

I. Think About It

❶ What kinds of fungi have you observed? Where?

_____
_____
_____
_____

❷ Have you seen mold? Where and what did it look like?

_____
_____
_____
_____

❸ Have you seen a mushroom? Where and what did it look like?

_____
_____
_____
_____

❹ Have you seen yeast? Where and what did it look like?

_____
_____
_____
_____

❺ Have you observed molds on fruit or bread? If so, what did they look like?

_____
_____
_____
_____

❻ Have you observed molds on other objects? If so, what objects had molds and what did the molds look like?

_____
_____
_____
_____
_____
_____

II. Observe It

❶ Put a piece of newspaper or plastic in a warm, moist corner of the kitchen. Cut a piece of fruit in half and place the pieces on the newspaper or plastic. Then cover your work area with newspaper or plastic to keep mold from growing in other areas. Allow the fruit to sit for a few days until you see mold growing.

❷ When the fruit is moldy, take 3 plastic bags and label them: **Control**, **Moldy Fruit**, and **Moldy Fruit + Bread**. Write the date on each.

❸ Put 5 milliliters (1 teaspoon) of water in each bag.

❹ Put on disposable gloves to handle the moldy fruit.

❺ To the bag labeled **Control**, add a fresh square of bread and seal the bag.

❻ To the bag labeled **Moldy Fruit**, add one piece of the moldy fruit and seal the bag.

❼ To the bag labeled **Moldy Fruit + Bread**, add a fresh square of bread and a piece of the moldy fruit. Seal the bag.

❽ Leave the sealed bags at room temperature (20°-30° C, or 68°-85° F) in an area where they will be undisturbed. Wash your hands and clean up carefully!

❾ Check the samples daily for 3-7 days. **Do not open the bags to look at the samples.** In the following table, describe what the mold looks like and how much mold is growing on each sample. The warmer the environment, the faster the molds are likely to grow.

Description of Mold and Amount of Mold

	Control	Moldy Fruit	Moldy Fruit + Bread
Day 1			
Day 2			
Day 3			
Day 4			
Day 5			
Day 6			
Day 7			

⑩ After your experiment is over, throw the bags away **without opening them!**

III. What Did You Discover?

❶ Did mold grow on the bread sample that was placed in the bag with the moldy fruit?

_____

_____

_____

❷ If so, were the molds the same color? Did they look different?

_____

_____

_____

❸ Did any mold grow on the **Control**? If so, what color was it?

_____

_____

_____

❹ Did you observe any other changes in the **Control** (bread only)? In the **Moldy Fruit**? In the **Bread + Moldy Fruit**? If so, describe them.

_____

_____

_____

_____

IV. Why?

Mold spores are found everywhere. Most food that is left out or stored in an unsealed container will eventually get mold spores on it and show signs of mold growth.

In this experiment you used a "control." A control is a part of an experiment that does not undergo testing. Scientists can then take the parts of the experiment that were tested and compare them to the control that was not tested. In this way they can tell whether testing caused changes that did not happen to the control.

In this experiment you were testing bread to see if mold that grew on the fruit would grow on the bread. The piece of bread you placed in the bag by itself was your control for this experiment. If the **Control** bread has no mold on it and the bread in the **Bread + Moldy Fruit** bag does have mold, this means that the mold on the bread is likely the same mold as that on the fruit. However, if the control bread shows signs of mold, then it is more difficult to determine if bread in the **Bread + Moldy Fruit** bag got the mold from the fruit or if it would have gotten moldy anyway.

V. Just For Fun

Can bread mold grow on fruit? Reverse the experiment and see if bread mold will grow on a piece of fruit. Write or draw your observations below.

	Description of Mold and Amount of Mold		
	Control	Moldy Bread	Moldy Bread + Fruit
Day 1			
Day 2			
Day 3			
Day 4			
Day 5			
Day 6			
Day 7			

Experiment 9

Nature Walk: Observing Plants

Introduction

The first and most important part of learning about plants is to observe them. What can you notice about plants?

I. Think About It

❶ What different types of plants have you observed?

_____
_____
_____
_____

❷ Do you have plants that live in your house? Are they different from each other? Why or why not?

_____
_____
_____
_____

❸ Do you have plants that live in your yard? Are they different from each other? Why or why not?

_____
_____
_____
_____

❹ What kinds of flowers have you seen on plants? How were they different from each other?

_____

_____

_____

_____

❺ Have you ever seen grass and weeds? How do you tell which is which?

_____

_____

_____

_____

❻ Have you noticed that some plants grow in one place but not in another? What do you think is different about the places where plants grow?

_____

_____

_____

_____

II. Observe It

Scientists keep notebooks where they can write down observations they make while they're studying living things outdoors. These are called *field notebooks*. In this experiment you'll make your own field notebook.

❶ Take a notebook, a pencil, some colored pencils, and a camera if you have one, and put them in a backpack along with a snack and some water. Walk outside in a wooded area, a park, or in your yard—anywhere you can observe a variety of plants.

❷ Locate two or more different kinds of plants you would like to study.

❸ Draw the plants you see. Even if you are not a skilled artist, you can draw the basics of what the plants look like—for example, what size and shape the leaves are, whether there are few or lots of leaves, any flowers or seed pods or fruit they may have, how tall or short they are, and other features you observe. Also note the color of the plants.

❹ Notice where the plants are growing, what the soil looks like, how much sun they are getting, the temperature, and how much water they have. Write details such as these in your notebook along with any other observations you'd like to record.

❺ If you have a camera, take a photo, print it, and tape it in your notebook alongside your drawing.

❻ Take more nature walks and add new plant observations to your field notebook. Now you are a biologist!

III. What Did You Discover?

❶ Where were the plants you observed? In a field? In a pot? Near a river or stream?

_____

_____

❷ Did the plants have little, medium, or a lot of water?

_____

_____

❸ Did the plants have little, medium, or lots of sunshine?

_____

_____

❹ What was the surrounding temperature?

_____

_____

❺ How would you describe the kind of environment the plants you observed live in? (desert, wet or marshy, rocky, forest, etc.) What details did you observe?

_____

_____

_____

IV. Why?

Observing plants and the environment they live in is the first step to learning about plants.

Making careful observations allows you to learn things such as how tall or short a plant is, what color the leaves or flowers are, whether or not it has thorns, and if it is fuzzy or smooth. By noticing the features of different plants, you will become aware of the wide variety of plants and how in some ways they are the same as each other and in some ways they are different. Biologists group plants into divisions based on careful observations of the features the plants have.

By making careful observations, you can also notice that different plants require different surroundings, or environments, in order to grow and be healthy. Plants require air, water, sunlight, and soil that has nutrients in it. But some plants can only live where the temperatures stay warm and there is not too much water. Other plants need freezing temperatures in the winter. Some plants need to live in the shade and others need to live where they get lots of sunlight and there is no shade. Some plants grow best when people take care of them, and most do just fine on their own.

The study of plants is called *botany*. Botanists have discovered many amazing things about plants and continue to make new discoveries.

V. Just For Fun

Record more observations in your field notebook. Choose two or more different kinds of plants that you'd like to observe over a period of several months. You may want to observe the same plants as you did for the *Observe It* section, or you may want to choose different plants.

Ask yourself questions about the plants and their environment. How do the plants change—or do they stay the same? Do changes in things such as temperature and amount of rainfall effect the plants? Are bugs or animals eating them? Do the plants change with the seasons? Asking questions leads to more and better observations.

Record your observations in your field notebook by drawing, writing about, and photographing the plants. You can also record anything else you find of interest in the plants' environment.

Experiment 10

Who Needs Light?

Experiment 10: Who Needs Light?

Introduction

What happens to plants when they don't get sunlight? Do this experiment to find out.

I. Think About It

❶ List three things a plant needs to have in order to live.

_____

_____

_____

❷ What do you think would happen to a plant if it did not get light from the Sun?

_____

_____

_____

_____

_____

_____

_____

II. Observe It

❶ Take two small plants that are the same kind and about the same size.

❷ Carefully observe each plant. Note any unique features they have.

❸ Make a list of words that describe the plants and their features.

Experiment 10: Who Needs Light? 95

❹ Label one plant **A** and the other **B**. Draw each plant.

A **B**

❺ Take the plant labeled **A** and put it in a sunny place.

❻ Take the plant labeled **B** and put it in a dark place.

❼ Describe what you think will happen to each plant.

Plant A

_____
_____
_____
_____
_____
_____

Plant B

_____
_____
_____
_____
_____
_____

❽ Make a schedule for watering your plants on a regular basis. Be sure to water both plants with the same amount of water.

Experiment 10: Who Needs Light? 97

Draw each plant after week 1.

Draw each plant after week _____

A   **B**

Experiment 10: Who Needs Light? 99

Draw each plant after week _____

Draw each plant after week _____

Experiment 10: Who Needs Light? 101

III. What Did You Discover?

❶ How did the plants look on the first day?

Plant A _____

_____

Plant B _____

_____

❷ How did the plants look after the first week?

Plant A _____

_____

Plant B _____

_____

❸ How did the plants look after the last week?

Plant A _____

_____

Plant B _____

_____

❹ Describe any differences you observed between the two plants.

Plant A _____

_____

Plant B _____

_____

IV. Why?

A regular houseplant needs sunlight to make food. If a houseplant is not able to get sunlight, it cannot make the food it needs to stay healthy. Eventually, a houseplant will die if it does not get enough sunlight.

When you put one plant in the dark and keep one plant in the sunlight, you are testing what happens to a plant that does not get sunlight. Why do you think you needed two plants—one in the sunlight and one in the dark?

You used two plants because, as a scientist, you want to make careful observations when you change something. When you use two plants (one in the sunlight and one in the dark), you can easily compare any changes in each of the plants. You want to know what happens to a plant that is in the dark compared to a similar plant that stays in the sunlight.

This is called using a *control*. The control shows you what will happen if nothing is changed. In this way, scientists can be sure that they will be able to observe what happens when something is changed. In this experiment, you observed what happened when sunlight was taken away from a plant. Your control plant (plant **A**) showed you what the plant looked like when it had sunlight. The plant you took away from the sunlight (plant **B**) showed you what happened to the plant when it could not use the Sun's energy to make food. Using a control helped you determine what happens when a plant does not get sunlight.

V. Just For Fun

Do another experiment using two more plants that are both the same kind and about the same size. This time water one plant and don't water the other one. Based on the last experiment, think of the steps you will need to take to preform this experiment. Record your results at the beginning and the end of the experiment.

Plant A — Beginning

Plant B — Beginning

Plant A — Ending

Plant B — Ending

Experiment 11

Thirsty Flowers

Introduction

How does water travel through a flower? This experiment will help you find out.

I. Think About It

❶ If you put a white carnation into a glass of colored water, what do you think will happen to the flower?

_____
_____
_____
_____

❷ Draw a picture showing what you think will happen.

II. Observe It

❶ Carefully observe the carnation. Draw your observations.

❷ Take one carnation and split it in half lengthwise. Draw your observations below.

❸ Take another carnation and place the stem in colored water. Observe what happens to the flower. Use the next boxes to draw what you see happening over the next several minutes.

After _____ minutes.

After _____ minutes.

After _____ minutes.

❹ Cut the stem open. Draw what you see.

III. What Did You Discover?

❶ What did the carnation look like before you added the colored water?

_____

_____

_____

❷ What did the carnation look like inside?

_____

_____

_____

❸ What happened to the carnation when you put it into the colored water?

_____

_____

_____

❹ What did you observe in the stem after it was in the colored water?

_____

_____

_____

IV. Why?

A carnation is a flower that usually has a long green stem. The stem is the part of the plant that carries water and nutrients up the plant. The roots of the plant take water and nutrients from the soil, and tissues in the stem then act like little straws that draw water from the roots up the plant.

When you place a carnation in colored water, you can observe what happens to the flower. Because the water is colored, you can watch the flowers "drink" the water from the glass. The colored water travels up to the top of the stem of the carnation, and when it reaches the flower, it starts to color the petals.

When you cut open the carnation stem, you can observe the tissues that move the water up the stem through the plant. With the stem cut open, you may be able to observe that the tissues in the stem are colored too.

These tissues in a carnation stem are designed to move the water one way—up the stem. The water does not come back down through the stem. (Do you think that is true? Why don't you try it—take the stem out of the colored water, and see if the colored liquid runs back out.)

A plant "drinks" water much like you drink water from a straw. Can you explain how a plant "drinks" water?

V. Just For Fun

Repeat the experiment with a different kind of white flower. Observe and record how long it takes for the flower to become colored. Did anything different happen with this flower than with the carnation? Record your observations below, noting similarities and differences.

Observations of Another Thirsty Flower

Experiment 12

Growing Seeds

Introduction

This experiment will help you find out what happens as a seed grows into a plant.

I. Think About It

❶ If you put a dried bean in a jar, add some water, and then let it sit for several days, what do you think will happen?

_____

_____

_____

_____

❷ Draw a picture showing what you think will happen.

II. Observe It

❶ Carefully observe a dried bean. Look at the outside. Draw your observations.

❷ Take the bean and split it in half lengthwise. Draw your observations.

❸ Take a clear glass jar and a piece of absorbent white paper. Cut a piece of the paper so that it is long enough to go all the way around the jar. Then wrap the piece of paper around the inside of the jar.

❹ Place two dried beans between the paper and the jar. The paper should hold the beans against the side of the jar, but if it doesn't, you can tape the beans to the jar.

Make sure the beans are not touching the bottom of the jar but are placed about 6-12 mm (1/4-1/2 inch) above the bottom.

❺ Pour some water in the bottom of the jar so that the water contacts the absorbent paper but not the beans.

❻ Place plastic wrap on top of the jar and fasten it with a rubber band to seal the jar and prevent evaporation of the water.

❼ In the box on the next page, draw the beans and the jar. Include any details you observe.

Observations: Day 1

❽ Check your beans frequently to observe their growth. As you notice changes, record your observations in the following boxes. Add water to the jar as needed to make sure the paper stays moist.

Observations: Day _____

Observations: Day _____

Observations: Day _____

Observations: Day _____

Observations: Day _____

III. What Did You Discover?

❶ What did the inside of the dried bean look like when you opened it? What did you find inside?

_____
_____
_____
_____

❷ How many days did it take for the beans to start growing?

_____

❸ Which part started to grow first? Which way did it grow—up or down?

_____
_____

❹ How many days did it take for the beans to turn into seedlings? In your own words, describe briefly how they grew from bean to seedling.

_____
_____
_____
_____
_____

IV. Why?

A bean is a seed. Seeds are how most plants begin. Inside a bean you can see the embryo that will grow into the little plant, or seedling. Inside the bean you can also see the food the embryo uses to grow until it has the roots and leaves it needs to make its own food.

When you put a bean in the ground, it will start to sprout. You can watch a bean sprout by putting it into a clear jar and adding water. The bean starts to sprout a root first. The root will grow downward, finding its way to the ground. A root knows which direction to grow, and it will not grow upward toward the Sun but down into the ground. The shoot of the plant will grow next. It will grow upward toward the Sun so that when the leaves come out, they can collect the sunlight.

The bean continues to grow the roots and the shoot until it becomes a seeding. When the seedling has leaves and a root big enough to gather nutrients, it no longer needs the food it had inside the seed. It is ready to become a big plant!

V. Just For Fun

Repeat the experiment using a different kind of seed. You might try a different type of dried bean or pea seed. You might try garden seeds you get at the store. Or you might gather seeds from a raw fruit or vegetable you are eating, such as a tomato, cucumber, watermelon, or squash. If you gather the seeds yourself, let them dry before putting them in the jar. Record your observations.

Observations of _____ Seeds

Experiment 13

Nature Walk: Observing Animals

Introduction

Why is observing animals the most important way to learn about them? Go on a nature walk to find out!

I. Think About It

❶ What types of animals have you observed?

_____
_____
_____
_____

❷ Do you have animals that live in your house? What are they and how would you describe them?

_____
_____
_____
_____

❸ Do you have animals that live in your yard? What are they and how would you describe them?

_____
_____
_____
_____

❹ Have you ever seen animals at a zoo? If so, what kinds? How would you describe them? How are they the same and how are they different?

_____

_____

_____

_____

_____

_____

❺ What features of animals have you observed that are common to many animals?

_____

_____

_____

_____

_____

❻ What unique features let you know the difference between the kinds of animals you are observing?

_____

_____

_____

_____

II. Observe It

❶ Take your field notebook, a pencil, and some colored pencils and put them in a backpack along with a snack and some water.

❷ Find a wooded area, park, or other place near your home where you can observe animals. Or you might take a trip to the zoo.

❸ Locate several animals you would like to study. Notice what features each animal has. Write the names of these animals and their features in your field notebook.

❹ Observe what each animal is doing. Are they moving, eating, sleeping, communicating, or watching you? Where are they—on the ground, in a tree, in a pond, or someplace else? What else can you observe? Write down your observations.

❺ Now make a quick drawing of each animal. You can record their shape and color, what their ears and tail look like, and any other features you notice.

❻ If you have a camera, take a photo of each animal, print it, and place it in your notebook alongside your drawing.

❼ Take more nature walks and add new animal observations to your field notebook.

III. What Did You Discover?

❶ Which animals did you observe? Where were they?

_____
_____
_____
_____
_____
_____

❷ Did the animals move? If so, how and how much?

_____
_____
_____
_____
_____
_____

❸ Did you observe any animals that did not move? If so, why were they not moving? Where were they?

_____
_____
_____
_____
_____

❹ Were any of the animals eating? If so, what were they eating and how were they eating it?

❺ Were any of the animals communicating? If so, how?

❻ What did you observe about animals that you hadn't noticed before?

IV. Why?

Observing animals in their environment is the first step to learning about them. By continuing to make careful observations, you will know what the different animals look like, what color they are, what their skin or hair or outer shell looks like, how large or small they are, and what kinds of food they eat. You can learn how different kinds of animals interact with each other and depend on each other.

Keeping a field notebook allows you to keep all your observations in one place so you can refer to them later. You can add more information about each animal as you make new observations, and you can add more kinds of animals to observe. When you observe animals over a long period of time, you learn about how they grow, how their activities change with the seasons, and how they interact, along with many other interesting details.

V. Just For Fun

Keep adding to your field notebook. Observe the animals in your area for several months. How do they change—or do they stay the same? Find some additional animals to observe. Write about, draw, and photograph them. (Did you remember that bugs are animals?)

Experiment 14

Red Light, Green Light

Introduction

Do you think snails and earthworms will move across any surface? Test it!

I. Think About It

❶ What do you think snails like?

___
___
___
___
___

❷ What do you think snails don't like?

___
___
___
___
___

❸ What do you think earthworms like?

___
___
___
___
___

❹ What do you think earthworms don't like?

___

❺ If you want to keep snails out of your garden, how would you do it?

___

❻ If you want to encourage earthworms to stay in your garden, how would you do it?

___

II. Observe It

❶ Take a large plastic box or tray and add enough garden dirt to cover the bottom.

❷ Take some water and moisten the garden dirt in one-half of the box (one end). Use just enough water to make the dirt moist. The other half of the dirt needs to stay somewhat dry.

❸ Perform a control experiment by placing the earthworms and/or snails on the dry side of the box. Observe whether or not they move to the moist end. Record your observations below.

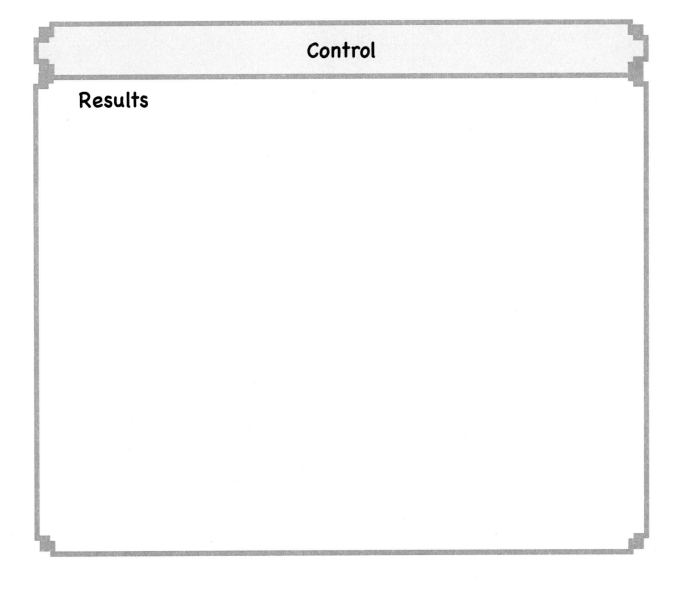

Control

Results

❹ Carefully remove the snails and/or earthworms and place them back in their holding box.

❺ Choose one of the powders and pour it in a line across the soil from one side of the box to the other about halfway between the two ends of the box.

❻ Place the earthworms and/or snails on the soil in the dry end of the box and observe whether or not they cross the powder barrier.

❼ Record your observations.

Powder Used _____

Results

Experiment 14: Red Light, Green Light 141

❽ Gently remove the snails and/or earthworms and place them back in their holding box.

❾ Remove the powder and fill in this area with new dirt, then pour a different powder in a line across the dirt. Repeat Steps ❻-❼

Powder Used _____

Results

⑩ Repeat Steps ❽-❾ using two different powders.

Powder Used _____

Results

Powder Used _____

Results

III. What Did You Discover?

❶ When there was no barrier, did the snails and/or earthworms move from the dry end to the moist end of the box? Why or why not?

___

❷ Were there any powders that the snails and/or earthworms would not cross? Why or why not? If there were any barriers they wouldn't cross, list them.

___

❸ Use the table below to chart which of the powder barriers the snails and/or earthworms crossed and which they would not cross.

Powder	Snails WOULD cross	Snails would NOT cross	Worms WOULD cross	Worms would NOT cross

❹ Could you use this experiment to find ways to keep snails out of your garden but earthworms in your garden? How would you do it?

_____
_____
_____
_____
_____

IV. Why?

Worms and snails are not as smart as octopuses, but they can choose good environments, and they have behaviors that help them escape from birds so they won't be eaten. They have the ability to know if some substances are dangerous to them and then to avoid those substances.

Snails can withdraw into their shell for protection, and they hide in dark moist places, like under rocks and foliage. Because snails need to stay moist in order to live, they usually hide during the day. Their bodies can dry out and die in the hot sun. Hiding during the day also protects them from predators. If the weather is cloudy or rainy, you may see snails out during the day, and they are often out in the cooler times of dawn and dusk and at night. When snails come out, they eat a lot! They can be very harmful to a garden, destroying many plants.

Earthworms are great to have in your garden because they help plants grow. Like snails, earthworms need to be in a moist environment or they will dry out and die. To escape predators, earthworms spend most of their time underground. You might see earthworms on the sidewalk after a big rain, but scientists are still not sure why earthworms come out when it rains. As a result of doing experiments, scientists think that earthworms communicate with each other through touch and that earthworms sometimes travel in groups.

V. Just For Fun

Do you think you can do a similar experiment to find out how to keep ants out of your house? Find an active anthill to use in your experiment.

On the next page write the steps you'll take to perform your experiment.

Ant Experiment

Perform your experiment and record your observations on the next page.

Results of Ant Experiment

Experiment 15

Butterflies Flutter By

Introduction

This experiment is about observing the life cycle of an arthropod. A *cycle* is a series of events that repeat, and a *life cycle* is the series of events from birth through death of an animal.

I. Think About It

❶ How did the butterfly get its name? Is it because butterflies eat butter? Or do you think it is that butterflies are sometimes yellow—like butter?

Write or draw how you think the butterfly got its name.

_____

_____

_____

_____

❷ How does a caterpillar turn into a butterfly? Draw or write how you think a caterpillar becomes a butterfly.

Experiment 15: Butterflies Flutter By 151

II. Observe It

❶ The Beginning: The Egg
Draw and/or write what you see.

_____
_____
_____
_____
_____
_____
_____
_____

❷ **The Middle: The Caterpillar**
Draw and/or write what you see.

Experiment 15: Butterflies Flutter By 153

❸ The Change: The Chrysalis
Draw and/or write what you see.

❹ **The End: The Butterfly**
Draw and/or write what you see.

❺ Draw the life cycle of the butterfly as you observed it.

❻ What else did you notice that was very interesting? Draw it below.

III. What Did You Discover?

❶ Did the butterfly eggs look like you thought they would? Why or why not?

_____
_____
_____
_____
_____

❷ Did the caterpillar look like you thought it would? Why or why not?

_____
_____
_____
_____
_____

❸ Were you able to observe exactly what was happening in the chrysalis? Why or why not?

_____
_____
_____
_____
_____

❹ Did the butterfly look the way you thought it would? Why or why not?

_____
_____
_____
_____
_____

❺ Describe your favorite part of the butterfly life cycle. Explain why it is your favorite part.

_____
_____
_____
_____
_____

IV. Why?

The butterfly is a creature that starts as something completely different—a caterpillar. If you had never watched a butterfly egg turn into a caterpillar and a caterpillar turn into a butterfly, you would not know that they are the same creature. It takes a keen eye and careful observation to find out what the life cycle is for the butterfly.

How did the butterfly get its name? You probably found out that no one is quite sure. There are ideas about how the butterfly got its name, but not everyone agrees. One idea is that the word butterfly comes from a very old word *buturfliog* which is a word coming from "butter" and "fly." But why butter? One idea is that butterflies like to eat butter and land on creamy, buttery foods in kitchens. The German word for butterfly means "milk thief," so maybe butterflies like butter and milk. But no one is really sure where the name for butterfly came from.

Sometimes disagreements happen even in science. We know much about the life of a butterfly because we can observe it. But scientists don't know everything because scientists can't observe everything. You were not able to observe exactly what was going on inside the chrysalis as the butterfly changed, because you couldn't see inside the chrysalis—but you were able to make some observations. The most important job of scientists is to make careful observations and then to record exactly what they see, even if they can't see everything—just like you did in this experiment!

V. Just For Fun

Using your *Student Textbook* as a reference, write some features of arthropods in your field notebook. Take your field notebook and go outside to look for arthropods. Use the features you have written down to help determine if the creature you're observing is an arthropod.

Also look for arthropods in different stages of their life cycle. Can you find eggs? Larvae? (Larvae are the newly hatched form of some insects. For example, caterpillars are larvae.) Chrysalises? Young arthropods? Adult arthropods?

In your field notebook, make notes and drawings about what you observe.

Experiment 16

Tadpoles to Frogs

Experiment 16: Tadpoles to Frogs

Introduction

Discover the different stages in a frog's life cycle.

I. Think About It

❶ How did the frog get its name? Is it because frogs sit on logs? Or do you think it is that frogs have deep voices? Write or draw how you think the frog got its name.

_____
_____
_____
_____
_____
_____

❷ What do you think will happen as a tadpole turns into a frog? Write or draw what you think.

II. Observe It

❶ The Beginning: The Egg
Draw and/or write what you see.

❷ **The Middle: The Tadpole Eating**
Draw and/or write what you see.

❸ The Change: The Tadpole with Hind Legs
Draw and/or write what you see.

❹ **The Change: The Tadpole with Front Legs**
Draw and/or write what you see.

❺ The End: The Adult Frog
Draw and/or write what you see.

❻ Draw the life cycle of the frog as you observed it.

❼ What did you notice that you thought was very interesting? Draw it below.

III. What Did You Discover?

❶ Did the frog eggs look like you thought they would? Why or why not?

_____
_____
_____
_____
_____
_____

❷ Did the tadpole look like you thought it would? Why or why not?

_____
_____
_____
_____
_____
_____

❸ Were you able to observe exactly what was happening when the tadpole started to change? Why or why not?

_____
_____
_____
_____
_____

❹ Did the adult frog look like you thought it would? Why or why not?

___

❺ Describe your favorite part of the frog life cycle. Explain why it is your favorite part.

___

IV. Why?

A frog starts life as an egg and then becomes a tadpole before it becomes an adult frog. It changes significantly during its life cycle. A tadpole doesn't look like a frog but more like a fish. Even though a tadpole may look like a fish, it is not a fish. If you never observed a tadpole changing into a frog, you wouldn't know they are the same creature.

Some creatures like frogs and butterflies undergo a drastic change in appearance when they become adults. This process is called *metamorphosis*. Metamorphosis simply means "to change form or shape."

Humans do not undergo a metamorphosis as they grow into adults. Although you will look different when you are an adult than you do now, you do not completely change your form or your shape as you grow. You may get taller and your body will change proportions, but overall you keep the same shape and form you had when you were born. You are a different kind of creature than a frog or a fish!

V. Just For Fun

Go outside and look for frogs and toads, lizards and snakes, and all kinds of birds. What can you observe about where they are in their life cycle? Do you see any eggs? Babies? Adults? How many different kinds of frogs, toads, lizards, and snakes can you see? If you're near water, look for fish too!

Take your field notebook with you and make notes and drawings about what you observe.

More REAL SCIENCE-4-KIDS Books
by Rebecca W. Keller, PhD

Building Blocks Series yearlong study program — each Student Textbook has accompanying Laboratory Notebook, Teacher's Manual, Lesson Plan, Study Notebook, Quizzes, and Graphics Package

Exploring Science Book K (Activity Book)
Exploring Science Book 1
Exploring Science Book 2
Exploring Science Book 3
Exploring Science Book 4
Exploring Science Book 5
Exploring Science Book 6
Exploring Science Book 7
Exploring Science Book 8

Focus On Series unit study program — each title has a Student Textbook with accompanying Laboratory Notebook, Teacher's Manual, Lesson Plan, Study Notebook, Quizzes, and Graphics Package

Focus On Elementary Chemistry
Focus On Elementary Biology
Focus On Elementary Physics
Focus On Elementary Geology
Focus On Elementary Astronomy

Focus On Middle School Chemistry
Focus On Middle School Biology
Focus On Middle School Physics
Focus On Middle School Geology
Focus On Middle School Astronomy

Focus On High School Chemistry

Super Simple Science Experiments

21 Super Simple Chemistry Experiments
21 Super Simple Biology Experiments
21 Super Simple Physics Experiments
21 Super Simple Geology Experiments
21 Super Simple Astronomy Experiments
101 Super Simple Science Experiments

Note: A few titles may still be in production.

Gravitas Publications Inc.
www.gravitaspublications.com
www.realscience4kids.com